NASA
WINGS

NASA WINGS

NIGEL MACKNIGHT

Published in 1992 by Osprey Publishing Limited
59 Grosvenor Street, London W1X 9DA.

© Nigel Macknight 1992

All rights reserved. Apart from any fair dealing for the purpose of private study, research, criticism or review, as permitted under the Copyright, Designs and Patents Act, no part of this publication may be reproduced, stored in a retrieval system, or transmitted in any form or by any means, electronic, electrical, chemical, mechanical, optical, photocopying, recording or otherwise, without prior written permission. All enquiries should be addressed to the Publishers.

British Library Cataloguing in Publication Data
Macknight, Nigel, 1955 —
    NASA Wings
    1. United States. NASA. Flight research aircraft. Illustrations
    I. Title II. Series
ISBN 1 85532 216 1

Editor Dennis Baldry
Page design Nigel Macknight
Printed in Hong Kong

Front cover Perhaps NASA's most famous aircraft are the two heavily-modified Boeing 747s used to ferry Space Shuttle Orbiter vehicles from place to place. Designated SCA – for Shuttle Carrier Aircraft – these gargantuan ex-airliners feature internal strengthening, large vertical fins at the tips of the horizontal stabilisers for additional directional stability, and support struts atop the upper fuselage

Back cover One of two General Dynamics F-16XLs employed in NASA's ongoing programme to improve laminar (smooth) airflow on aircraft flying at sustained supersonic speeds. It is the first programme to examine laminar-flow at speeds faster than sound. The resulting technological data will be available for the development of future high-speed aircraft, including commercial transports

Title page Graphically illustrating aviation's burgeoning technological advance and NASA's pivotal role in that process, one of three Lockheed SR-71 strategic reconnaissance aircraft loaned to the agency by the US Air Force for use in high-speed flight research overflies the Bell X-1E, sister ship of the X-1, first aircraft to fly faster than the speed of sound

Right One of NASA's two McDonnell Douglas AV-8 Harrier V/STOL aircraft hovers over the ramp at the agency's Ames facility at Moffett Field, near San Francisco

For a catalogue of all books published by Osprey Aerospace please write to:

The Marketing Department, Octopus Illustrated Books,
1st Floor, Michelin House, 81 Fulham Road, London SW3 6RB

Introduction

With its necessarily high-profile involvement in space exploration, NASA is widely perceived as an organisation solely devoted to activities beyond Earth's atmosphere. Hopefully, this book will play a small part in generating wider recognition of the agency's unrelenting commitment to aeronautical research, which has been of inestimable benefit to both civil and military aviation. Over the years, NASA has test-flown a bewildering variety of aircraft types. They have ranged from modified versions of production aircraft to exotic one-of-a-kind research vehicles. This book is devoted to aircraft currently in NASA service. Many readers will be surprised to count no less than 50 aircraft types in the Inventory that appears at the back of this book. A key feature of *NASA Wings* is that every single aircraft type has been illustrated.

I'd like to point out that, in focusing on the aircraft themselves, we have presented only one facet of NASA's aeronautical effort. Bear in mind that a vast effort takes place on the ground, not only towards keeping the aircraft flying, but in conducting windtunnel and computational fluid dynamics (CFD) research programmes, aerostructures materials development, and all manner of hardware and software test programmes and flight-simulation activities.

Sifting through mountains of photographs to assemble a representative selection, then attempting to pen concise descriptions to accompany them has been as pleasant a task as any I've undertaken in 20 years of aviation writing. It's been an opportunity to continue many long-term working relationships with NASA personnel, and to begin a few new ones, and for that I'm grateful to Dennis Baldry and Tony Holmes of Osprey. I also wish to thank the following individuals for their invaluable contribution to the preparation of *NASA Wings*: Paul Wilson of P and S; Don James, Mike Landis and Mike Mewhinney at NASA/Ames; Ted Ayers, Jenny Baer-Reidhart, Denis Bessette, Lisa Bjarke, Marta Bohn-Meyer, Roy Bryant, Bill Dana, Gordon Fullerton, Don Gatlin, Don Haley, Steve Ishmael, Jack Kolf, Nancy Lovato, Dave Lux, Laurel Mann, Tom McMurtry, Ed Schneider, Rogers Smith, Lou Steers, Jim Stewart, and Milt Thompson at NASA/Edwards; Tom Jaqua and Dave Steitz at NASA Headquarters; Joe Algranty, Ace Beale and Frank Marlow at NASA/Ellington; Keith Henry, John Molloy and Harry Verstynen at NASA/Langley; Mike Gentry, Lisa Vazquez and Eileen Walsh at NASA/JSC; Linda Ellis at NASA/Lewis; Kelly Smith at NASA/Stennis; and Keith Koehler and Roger Navarro at NASA/Wallops.

The author Nigel Macknight was born in Corbridge, Northumberland, England in August 1955. He has translated a childhood enthusiasm for aviation into a successful career as a writer and broadcaster on a wide variety of aerospace topics, having turned professional in February 1977. In 1985 he founded *Spaceflight News* magazine and served as its Editor/Publisher until 1991. In addition to *Spaceflight News* and its progenitor, the *Shuttle Story* partwork, Macknight's work has appeared in *Aircraft Illustrated*, *Air Extra*, *Air Forces Monthly*, *Air International*, *Air Pictorial*, *Airshow*, *Aviation News*, *European Business Air News*, *Flypast*, and *Wings*. He has also authored over 100 articles on non-aerospace topics. In 1988, Macknight received an Aviation/Space Writers Association award for his book *Shuttle*, sales of which have exceeded 50,000 worldwide. *NASA Wings* is his fifth book.

Nigel Macknight is pictured above with NASA test pilot Ed Schneider after their high-performance flight in a privately-owned F-104 Starfighter over the Gulf of Mexico.

Contents

NASA/Ames 8
NASA/Edwards 30
Splash! 88
NASA/Ellington 90
NASA/Langley 108
NASA/Lewis 118
NASA/Wallops Island 120
NASA/Elsewhere 126
Inventory 127

This mural, painted by Robert T McCall, is displayed in the Visitor Center at NASA's Dryden Flight Research Facility at Edwards, California. It provides visitors with a thought-provoking composite view of how aerospace research and technology has advanced there through the years. Surrounding the world's first supersonic aircraft, the Bell X-1, at centre — and starting clockwise at 12:00 — are: the hypersonic North American X-15 (also depicted with its NB-52 carrier aircraft); the Douglas D-558-2 Skyrocket, the first aircraft to exceed twice the speed of sound, being air-launched from a Boeing B-29; the Vought F-8 Crusader which NASA equipped with the first supercritical wing; the Douglas X-3 Stiletto; the Parasev space-capsule recovery system; the 'wingless' Martin Marietta X-24B and Northrop M2 manned lifting bodies; the D-558-1 Skystreak; the XF-92A, first delta-wing aircraft; the Northrop HL-10 manned lifting body; the Bell-built Lunar Landing Training Vehicle; a subscale F-15 fighter Remotely-Piloted Research Vehicle (RPV) being dropped from a Sikorsky helicopter; the Bell X-5 variable-sweep-wing aircraft; the tailless Northrop X-4; the Space Shuttle Orbiter prototype, *Enterprise*, atop its modified Boeing 747 carrier aircraft, the SCA; the triple-sonic Lockheed YF-12 Blackbird; and the mighty North American XB-70 Valkyrie supersonic bomber prototype

NASA/Ames

Ames Research Center is located at Moffett Field, about 20 miles south of San Francisco. It was founded in 1940 as an aeronautical research laboratory and became part of NASA's predecessor organisation, NACA, in 1958. The centre's major research activities are computational and experimental aerodynamics, hypersonic aircraft studies, aeronautical and space human factors, life-sciences, space science, Solar System exploration and infrared astronomy. Ames boasts the world's largest windtunnel facility

Taken in 1986, this photo illustrates the diversity of aircraft operated out of NASA/Ames. Forming the line on the left, starting with the aircraft nearest the camera, are the Lockheed Jetstar, an F-104, the RSRA, the QSRA, the C-130, a T-38 and the DC-8 (the Jetstar, F-104 and RSRA are no longer at Moffett Field). On the right are the YO-3A, the King Air 200, the SH-3G, an ER-2, the Learjet 24 and the C-141 'Kuiper' observatory. In the centre, clockwise from 12:00, is the CH-47 (since withdrawn from service), the YAH-8B, the UH-1H an XV-15, an OH-58 (since withdrawn) and the NAH-1S

Named for the distinguished American astronomer Gerald P Kuiper, NASA's Lockheed C-141 is a modified version of the Starlifter military transport. It has been outfitted as a 'flying observatory', capable of climbing to 45,000 ft – above earthly smog and light-pollution – to ensure that astrophysicists on board get a clear view of the heavens

Above and right 'Kuiper' is equipped with a 36-inch infrared telescope (just forward of the wing) and an impressive suite of tracking, detection, observation and recording equipment. It is often deployed to remote sites around the world to obtain the best possible observing conditions. Targets of observation over the years have included solar-flares, supernovas, asteroids and Halley's Comet. When they set star-trackers to lock onto a particular constellation, astronomers at these consoles help 'fly' the C-141 via a direct link to the flight control system

This breathtaking view of a Space Shuttle Orbiter during re-entry was captured by the 'Kuiper' flying observatory in April 1981, on the reusable spacecraft's first orbital mission. Shuttle programme engineers wanted to compare the Orbiter's thermal profile with theoretical calculations, so the C-141 — deployed to Hickam AFB, Hawaii — was flown about 200,000 feet beneath *Columbia*'s flightpath as the spacecraft hurtled back into the upper reaches of the atmosphere at Mach 25. The aircraft's 36-inch infrared telescope, linked to a precision tracking apparatus, produced the requisite half-image

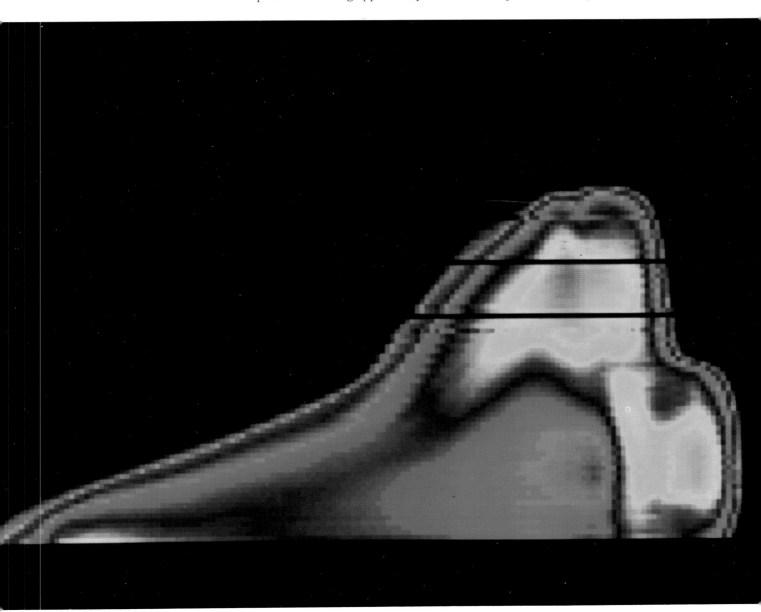

Above NASA's Lockheed C-130 Hercules is a modified version of the ubiquitous transport aircraft, crammed with sensors and data-recording equipment. Scientists place instruments aboard this aircraft to perform Earth-observations: viewing the oceans as well as land masses. 'Ground true' studies are also conducted, assisting cartographers (map-makers). Occasionally, new instruments are flown aboard the C-130 prior to being installed aboard Earth-observation satellites. This way, they can be thoroughly tested, yet can be returned for repair or recalibration if anything's amiss — a difficult task once a satellite's been launched!

Right The NASA C-130's belly is laden with sensors

Left High over the Golden Gate Bridge, with downtown San Francisco in the background, are the distinctive Lockheed TR-1 (foreground) and ER-2, both derived from the legendary U-2 strategic reconnaissance aircraft. NASA's examples perform strictly *overt* operations, mapping the globe and undertaking a wide variety of other Earth-observations duties, including mapping geological formations, vegetation classes and snow cover. High-altitude research tasks are also assigned to the two ER-2s and single TR-1 in NASA's charge. With a maximum ceiling of 72,000 feet, the aircraft can traverse the lower reaches of the stratosphere. In early-1989, an ER-2 and NASA's DC-8 were deployed to Stavanger in Norway to conduct ozone-depletion studies

Below A ground handler supports an ER-2's left wingtip as it rolls to a standstill, while a colleague wheels two dollies into place. The dollies will permit the ER-2 to taxi in at slow speed on its tandem landing gear

Right NASA research pilot Jim Burrilleaux clambers aboard an ER-2

Below The sole TR-1 and both ER-2s grace the Ames ramp

Above Originally built for use in Vietnam, the Lockheed YO-3A has a very high-aspect-ratio wing and was optimized for extremely quiet operation, in keeping with its original clandestine surveillance mission. NASA employs it to collect 'noise signal' data during tests of new rotorcraft blade designs. It is so quiet that it can be flown with acoustic-measurement devices mounted on booms extending from the wings and tailfin, without actually detecting its own engine noise. This way, it gathers 'uncorrupted' noise data, free from the usual distorting influences of ambient noise and Doppler-effect. The Sikorsky SH-3G Sea King is employed as a research platform for the Global Positioning System (GPS), a satellite-based high-accuracy military/civil navigation infrastructure. In addition, '735' serves as a short-field beacon and microwave landing system test platform and acts as a 'target' in studies of rotorcraft acoustic characteristics by flying in formation with the Ames YO-3A, as seen here

Right Close-up view of the SH-3G's interchangeable nose-mounted instrumentation ensemble

This derivative of Bell's famous HueyCobra gunship is employed on limited-vision-operations research. Designated NAH-1S, the distinctive slimline 'helo' is fitted with state-of-the-art laser and infrared guidance systems which provide the pilot with synthetic vision for safe flight operations in conditions of poor visibility. Interactive speech systems have also been tested

NASA/Ames has two McDonnell Douglas AV-8 Harriers. They are employed for some aerodynamic research work (into the basic handling characteristics of V/STOL aircraft), but their primary function is to support studies of future V/STOL aircraft design requirements. NASA researchers are particularly interested in the control laws required for the translation from vertical to horizontal flight and vice versa

This heavily-modified de Havilland Canada C8A Buffalo has been optimized for research into STOL (short take-off and landing) flight characteristics. Designated QSRA – for Quiet Short-haul Research Aircraft – it retains the original fuselage but totes a special Boeing-developed wing and AVCO-Lycoming YF-102 turbofan engines. The thrust from these is blown over the top surface of the wings and flaps to augment lift, conferring exceptional STOL performance and shielding engine noise from people on the ground. It is pictured during trials aboard the USS *Kitty Hawk* off San Diego during a joint US Navy/NASA evaluation exercise. A total of 37 touch-and-go landings and 16 full-stop landings were made. Arresting gear was not required, due to the QSRA's slow (65 knots) approach speed. A recent refinement is the incorporation of a pneumatic nosegear-extension device, which creates a rapid pitch-up during the take-off roll and further enhances STOL performance. NASA estimates that a typical 95-passenger aircraft based on QSRA technology would require only one-seventh the operating area, including runway and clearance airspace, required by current commercial airliners

This Douglas DC-8-72, retrofitted with upgraded CFM56 turbofan engines, serves as a medium-altitude atmospheric research platform. The former jetliner totes wing-mounted spectrometers and laser instruments, and has recently supported ozone-depletion studies in the Arctic and Antarctic regions. Ports set into the upper fuselage facilitate atmospheric observations

There are two Sikorsky UH-60 Blackhawks at NASA/Ames. The UH-60 illustrated here has pressure transducers, accelerometers and strain gauges attached to its rotor-blades to gather data on rotor airloads. The data is telemetered to an on-board data-acquisition system, then relayed to a ground-based receiver. The other UH-60 is designated the Rotorcraft Aircrew Systems Concepts Airborne Simulator, or RASCAL, and is currently validating a variety of software programmes used by NASA's Vertical Motion Simulator (VMS), under a joint US Army/NASA research effort that civilian industry is also keenly interested in. The VMS is a unique facility, capable of providing the equivalent of six storeys of vertical motion to simulate the flying charactertistics of helicopters, V/STOL and tilt-rotor aircraft, and even the Space Shuttle Orbiter in its approach and landing regime

With its ability to combine the attributes of a helicopter and a medium-sized fixed-wing transport aircraft, the Bell XV-15 has been the subject of intensive NASA scrutiny for many years. The agency actually has two of these exotic tilt-rotor research vehicles on strength. One serves at Ames, the other is on loan to Bell and based at that company's Fort Worth, Texas facility. A set of high-technology (all-composite) rotor blades have recently been fitted to the Ames XV-15 and flight testing has resumed after an extended hiatus

Right Beech King Air 200 N701NA is occasionally employed as a chase-plane, but is more often used to transport Ames personnel and their equipment from place to place. It joined the Ames inventory in August 1983

Below This Gates Learjet 24, tail number '705', has a small port set into one side of the fuselage, to facilitate telescopic observations. The Ames Learjet is equipped to track and image rocket launches and orbiting spacecraft, and has served as a chase aircraft during air-launches of the Pegasus booster from NASA's Boeing NB-52

(NASA/Ames also has a Bell UH-1H Iroquois, mostly used for chase work during flight tests of helicopters and the XV-15 tilt-rotor aircraft, and two Northrop T-38 Talons, used primarily for pilot proficiency-maintenance)

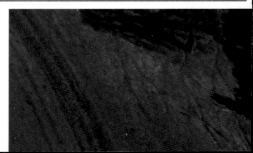

NASA/Edwards

Edwards AFB, located 100 miles north of Los Angeles, is the home of NASA's Dryden Flight Research Facility. It is acknowledged throughout the world as a test-flying mecca. The first supersonic flight was flown here in 1947, immortalising Chuck Yeager. Since then, the hypersonic X-15, the 'wingless' Manned Lifting Bodies, and dozens more radical flying testbeds have been put through their paces in the hallowed airspace over Mojave Desert

Below An aerial view of NASA's Dryden Flight Research Facility, showing its proximity to Rogers Dry Lake and other portions of Edwards AFB. The distinctive circular pattern on the lakebed is a Compass Rose and is marked with accurate compass headings. At left are markings for the lakebed runways used for Space Shuttle landings. At the centre of the picture is the main NASA hangar and administrative complex. The area to the right, connected by the taxiway extension, is the Shuttle processing area and the Mate-Demate Device, used to mount the Orbiters atop the Boeing 747 carrier aircraft for ferry flights back to the Kennedy Space Center, Florida launch site

Left A variety of types representative of NASA's fleet of research and support aircraft operated out of the agency's Edwards facility were displayed on the ramp in 1990. They were, from the front, left to right: the F-18 HARV high-alpha research aircraft; the second X-29 forward-swept-wing technology demonstrator; the F-15 HiDEC integrated-control research aircraft; the F-16XL laminar-flow research aircraft; three F-18 research support aircraft; a T-38 research support aircraft; an F-104 research aircraft; and, centre-rear, one of three SR-71s expected to fly in a future high-speed research programme. At far left is NASA's NB-52 air-launch aircraft with one of its payloads, a Pegasus space booster, on a trailer nearby. At far right is a Boeing 747 SCA Shuttle-carrier

With its wings on 'backwards', the Grumman X-29 is one of the most radical designs in aviation history. The second of two X-29s built is being flown by test pilots at Edwards to investigate a wide range of advanced concepts and technologies, providing an engineering database that could be used in the design and development of future aircraft. The first X-29 completed its test programme in 1988 and will not fly again. Foremost among the concepts under evaluation are the widespread use of advanced composite materials in aircraft construction, variable-camber wing surfaces, the unique forward-swept wing and its thin supercritical airfoil, strake flaps, and a computerized fly-by-wire flight control system that overcomes the aircraft's inherent instability. The X-29's forward-swept wing has been shown to reduce drag by up to 20 per cent in the transonic range. Furthermore, because the air moving over the forward-swept wing flows inward, rather than outward as it does on a conventional rearward-swept wing, the wingtips remain unstalled at higher angles of attack — a useful attribute for future fighter aircraft. The first attempt at designing an aircraft with a forward-swept wing was made by the Germans during World War 2. The concept proved unsuitable at that time, because the technology and materials did not exist then to construct a wing rigid enough to overcome bending and twisting forces without making the aircraft too heavy. The advent of composite materials in the 1970s opened new possibilities with strong, yet lightweight, structures

A rare view of X-29 No 2 arriving at Edwards on 7 November 1988, ensconced in a protective plasticised covering. The aircraft travelled by ship from Grumman's plant on Long Island through the Panama Canal to Port Hueneme and then was trucked to 'Eddie'

Above Condensation from the jet engine creates a sensational visual indication of the X-29's flightpath relative to its own orientation as its pilot undertakes high angle of attack (high alpha) tests with the unique aircraft. NASA, USAF and Grumman test pilots report that the X-29 has excellent control responses up to 45 degrees angle of attack and still has limited controllability at 66 degrees! This, without leading-edge flaps on the wings for additional lift and without moveable vanes on its engine exhaust to vector the direction of thrust

Right X-29 No 2 differs from the No 1 aircraft (not illustrated, as it is no longer operational) in having an anti-spin parachute installed just over the engine exhaust nozzle

Above Spectacular arrival of one of three Lockheed SR-71 Blackbird strategic reconnaissance aircraft loaned to NASA by the USAF in 1990/91 after the latter withdrew the type from operational duties. The single SR-71B is currently flight worthy, while the two A-model aircraft are being maintained at Edwards in 'flyable storage' status while NASA assesses research opportunities and experimentation that can benefit from the availability of a triple-sonic flying testbed. The agency operated YF-12s, predecessor to the SR-71, between 1969 and 1979, gaining much useful data on the structures and stability and control characteristics of air-breathing aircraft at high speeds and altitudes. A YF-12 was drafted in hurriedly to support studies of pilot-induced oscillation (PIO), following Shuttle landing difficulties

Right NASA research pilots Rogers Smith (in cockpit) and Steve Ishmael have been selected to perform piloting duties when the SR-71s commence their high-speed, high-altitude flight research programme. Data gathered will be applied to the design of future civil and military aircraft, including the X-30/National Aero-Space Plane (NASP)

Left NASA's first SR-71 enters the landing pattern in full reheat in February 1990

Below Drogue deployment prior to full development of the SR-71's drag-chute

Two General Dynamics F-16XLs are being employed in a programme to help improve laminar (smooth) airflow on aircraft flying at sustained supersonic speeds. It is the first programme to examine laminar-flow at speeds faster than sound. The resulting technological data will be available for the development of future high-speed aircraft, including commercial transports. The F-16XLs are being used because the unique 'cranked arrow' wing design is representative of the type of wing that will probably be used on future supersonic-cruise aircraft

Above The single-seat No 1 XL aircraft, a futuristic shape over the California desert. Clearly visible on the upper surface of the left wing is the thin experimental wing section, designed by Rockwell International and containing an active suction system. This device siphons off a portion of the drag-inducing layer of turbulent air through millions of tiny laser-cut holes, creating a smooth airflow over that section of wing

Right The twin-place No 2 XL pictured after its ferry flight from General Dynamics' Fort Worth, Texas facility to NASA's Edwards flight test centre in 1991. The aircraft will be modified to incorporate a different experimental wing surface to that fitted to the No 1 aircraft

Above NASA's Piper PA-30 Twin Comanche cruises low over the edge of Rogers Dry Lake at Edwards. It has been employed as a flying testbed in development of control systems for remotely-piloted vehicles (RPVs), a concept which led to highly-successful programmes such as the high-manoeuvrability HiMAT vehicle and a subscale F-15 test article. Over the years, '808' has also been used for spin and stall research related to the potentially dangerous wake-vortices formed behind large jetliners

Left Pictured with its engine in the extended position, this PIK 20E sailplane has been used to study high-lift aerodynamics and laminar flow on high-lift airfoils. Note the black dye patterns on the left wing, created for flow-visualisation purposes

Left A gorgeous study of NASA's externally-standard, but internally much-modified McDonnell Douglas F-15. It is pioneering a whole host of new technologies aimed at obtaining optimum aircraft performance. The system NASA has developed to investigate and demonstrate this capability is called HiDEC, for Highly Integrated Digital Electronic Control. Integrated digital electronic flight and engine control systems have resulted in improved rates of climb, fuel savings and increased engine thrust. The HiDEC aircraft has also been used to test and evaluate a 'self-repairing' flight control system capability. The highly-computerised system detects damaged or failed flight control surfaces, then automatically reconfigures the undamaged flight surfaces to compensate, allowing the mission to continue, or facilitating a safe landing. Nearly all the research being carried out in the HiDEC programme is applicable to future civilian and military aircraft

Below Flashback to the late-1970s, when NASA's F-15 was used to test the Space Shuttle's thermal protection tiles. Tiles on the F-15's right wing simulated those on the leading edge of the Shuttle Orbiter's wing, while tiles on the left wing represented tiles at the junction of the Orbiter's wings and fuselage

The HiDEC F-15, in full reheat, charges into the heat-haze at Edwards

NASA currently operates two Lockheed F-104 Starfighters, a single-seater and a two-seater. Although they were previously also employed as chase-planes, today they serve purely as research vehicles in their own right. Here, the single-seater is pictured flying into a water spray pattern being released from a tanker aircraft. This was a research flight to expose samples of Shuttle thermal protection system (TPS) materials to a variety of flight conditions, including rain and moisture. The samples are mounted on the unique, NASA-developed underfuselage Flight Test Fixture (FTF)

Close-up view of '826' revealing the FTF, just forward of the F-104's ventral fin. Research instrumentation is also mounted on the right wing pylon here. In the early days of the Space Shuttle programme, samples of the reusable spacecraft's thermal protection tiles were mounted on the wing flaps of '826', as well as on the FTF. This way, the tiles were subjected to a representative aerodynamic environment. At present, '826' is being employed to assess the flightworthiness, accuracy and reliability of a device known as the optical air-data system, or OADS. This prototype employs three pairs of small laser-light sheets to determine the speed and direction of atmospheric particles passing between them, thereby indicating aircraft speed, angle of attack and sideslip. The resulting data are processed on-board in realtime and telemetered to monitors on the ground. Operational systems, of course, will relay the data direct to the pilot. The OADS concept is being studied because it is non-intrusive and does not require orifices or other high-maintenance fixtures. It is also proving to be much more accurate than conventional air-data systems. A second laser system, still in the planning stages, is expected to be evaluated at Dryden in the future. It employs the Doppler frequency-shift method of measuring atmospheric particle velocity

NASA's two-seat F-104, '825', is used for pilot proficiency and training as well as for research. Among its most recent activities were flights to study X-30/NASP cockpit visibility concepts, including a periscope device

This Convair 990, formerly employed as a medium-altitude research platform at NASA's Ames facility near San Francisco, is now resident at Edwards. It is being converted to a landing systems research aircraft to test Space Shuttle landing gear and braking systems as part of the continuing effort to upgrade and enhance the reusable spacecraft's landing capabilities. A Shuttle landing gear retraction system is being installed in the lower fuselage area, between the CV-990's main landing gears. During the tests — which will be conducted at Edwards and at the Shuttle landing sites at White Sands, New Mexico and Kennedy Space Center, Florida — the Shuttle landing gear will be lowered immediately after the CV-990's three landing gears have contacted the runway, at the typical Shuttle landing speed of 230 mph. The tests will allow engineers to assess and document the performance of the Shuttle's main and nose landing gear systems, tyres and wheel assemblies and the nosewheel steering performance, and to know what to expect if the astronauts experience a tyre failure or related anomaly during an actual Shuttle touchdown. A thick blanket of hard rubber and Kevlar composite fibres is being attached to the CV-990's fuselage underside and inboard wing areas to protect it from any debris generated by blown Shuttle tires or other component failures. This particular CV-990 was built in 1962 and was operated by American Airlines and Modern Air Transport until acquired by NASA in 1975

A General Dynamics F-16 serves as a flying testbed for the Advanced Fighter Technology Integration (AFTI) programme, which is integrating and demonstrating new technologies for the next-generation close air support and battlefield air support (CAS/BAI) aircraft. Early in the joint NASA/Air Force/Army/ Navy programme, the AFTI F-16 demonstrated voice-activated commands, helmet-mounted sights, and flat turns and selective fuselage pointing utilising forward-mounted canards and a triplex digital flight control computer system

Above Close-up view of the AFTI F-16 over 'Eddie', highlighting the forward-mounted canard control surfaces, which permit no-bank turns. The present CAS/BAI phase of research will develop and demonstrate advanced technologies with emphasis on first-pass target acquisition and kill, survivable target area ingress and egress, and extension of the current capabilities into night operations in adverse weather while utilising manoeuvrability and speed at low altitude. The aircraft is equipped for automatic terrain following/terrain avoidance, ground collision avoidance and recovery – in the event of pilot disorientation and/or G-induced loss of consciousness (GLOC) – enhanced terrain-reference navigation, and task automation. It also totes a helmet-steered forward-looking infrared radar, and sensors that give the pilot a look-into-turn capability

Right Air-to-air refuelling of the AFTI F-16 by a USAF KC-10 tanker

NASA's Boeing NB-52 is the oldest example of the famous eight-engined bomber still flying, and has contributed to some of the most significant projects in aerospace history. Built in 1952, '008' was the eighth B-52 to come off the Boeing assembly line and was a USAF test aircraft for four years before it was assigned to NASA and fitted with a special wing-pylon, a launch panel operator's console and other modifications for the X-15 rocketplane programme. Two NB-52s air-launched the three X-15s over a ten-year period from June 1959 to October 1968. Highlights of the 199-flight X-15 adventure included the establishment of speed and altitude records of 4520 mph (Mach 6.7) and 354,200 feet. Subsequently, the NB-52s supported the Manned Lifting Body programme and drop-tested various RPVs. Only one NB-52 remains in active service today. Its most exotic payload these days is the Pegasus orbital booster, which can be seen here slung from the Boeing's starboard wing

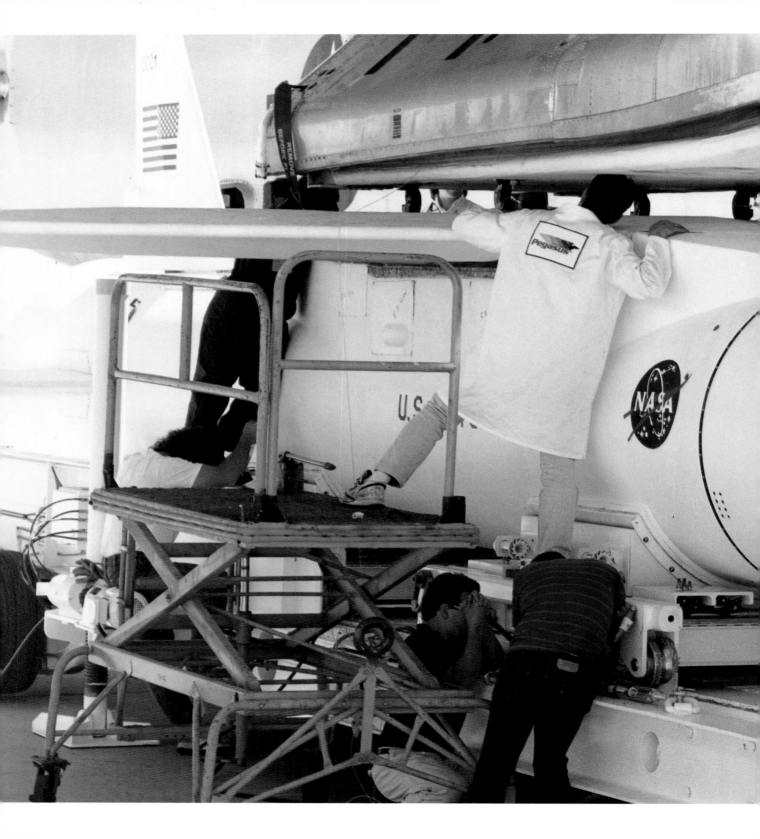

Pegasus was developed by Orbital Sciences Corporation and Hercules Aerospace, with support from NASA, DARPA and the USAF. A three-stage, all-solid-propellant booster, it is the first all-new unmanned launch vehicle to be produced by the United States for 20 years. Interestingly, Pegasus owes a portion of its parentage to the X-15, data gathered during the X-15 programme being used by the Pegasus design team. These close-up views reveal details of the NB-52's underwing pylon, previously used by the X-15 and the Manned Lifting Bodies. Pegasus has its own purpose-made adapter, with four pickup points. It serves as an interface with the NB-52 pylon, which has three pickup points

Following its conventional runway take-off, the NB-52 climbs out to the deployment area, some 60 miles off the California coast at Monterey, and establishes an altitude of 43,000 feet. Ten minutes before Pegasus is deployed, the NB-52 flight crew performs a series of S-turns for inertial navigation updating purposes

When the hooks are activated (by high-pressure gas), releasing Pegasus from the NB-52, the booster drops 300 feet in five seconds, before ignition of the first-stage rocket motor and initiation of the pitch-up sequence for the climb towards orbit. First-stage burnout occurs 81 seconds later, at an altitude of 208,000 feet and a speed of Mach 8.7. Twelve minutes after deployment from the NB-52, with second- and third-stage burns completed, Pegasus's satellite payload reaches Earth orbit

Right As these words were written, NASA was beginning a fourth series of drop-tests with an F-111 crew escape capsule test article for the USAF

Below right Flashback to 1977, when '008' was used to air-drop a Shuttle solid rocket booster parachute recovery system test article

Below Plans to retrofit the Space Shuttle fleet with drag-chutes could not progress until a series of deployment tests had been conducted using the NB-52. A total of eight test deployments were made in 1990 at landing speeds ranging from 160 to 230 mph. Former 'Shuttlenaut' Gordon Fullerton and test pilot Ed Schneider commanded the Boeing throughout the tests. When installed aboard the Shuttle Orbiters, the drag-chutes will reduce tyre and brake wear and shorten the reusable spacecraft's landing roll

The High Alpha Research Vehicle, or HARV, is a modified pre-production (the sixth built) McDonnell Douglas F-18 on loan to NASA from the US Navy. It is engaged in studies to validate computer codes and windtunnel research pertaining to manoeuvrability at high angles of attack (high alpha). In the first phase of research which began in mid-1987, 101 flights were made at angles of attack up to 55 degrees. The second phase of flight research began in 1991 and is utilising a thrust-vectoring system, permitting angles of attack near to 70 degrees. Phase Three will commence in 1994 and involve flight-testing moveable strakes on the forward fuselage. They will extend automatically at high angles of attack to interact with the strong vortices generated along the nose and produce large side forces for lateral control. The HARV F-18 is affectionately known as the *Silk Purse*, a reference to the fact that the aircraft began its life with NASA as the proverbial sow's ear — a basketcase left to languish at the US Navy's Patuxent River, Maryland facility

Initially, the HARV F-18 served in its original US Navy test paint scheme, primarily gloss white. This has since given way to a striking flat black scheme over the aircraft's upper surfaces, to increase the visibility of the white smoke trails released from the forward fuselage during 'off-surface' airflow behavior studies. 'On-surface' flow-visualisation can also be accomplished, by analysis of the patterns created by a special red liquid emitted from 500 tiny holes in the aircraft's nose cone (the liquid is poly-glycol methyl-ether, or PGME, mixed with red dye). The HARV has also been 'tufted', as shown on the preceding spread, to aid airflow visualization: short pieces of yarn were taped to the upper fuselage, leading edge extensions (LEXs) and wings. The HARV has five on-board cameras (still and video) to document airflow behaviour. Infrared film has been used to aid visualization of weak smoke patterns

The business end of the HARV's thrust-vectoring system comprises three hydraulically-actuated plates (turning vanes) made of Inconel metal. They provide both pitch and yaw forces to enhance manoeuvrability when the aerodynamic controls are either unusable or less effective than desired

Perhaps NASA's most famous aircraft is the heavily-modified Boeing 747 used to ferry Space Shuttle Orbiter vehicles from place to place. Designated SCA – for Shuttle Carrier Aircraft – there are, in fact, now two of these gargantuan ex-airliners in NASA service. The SCAs have been placed in the 'Edwards' section for continuity, since it is from here that they conduct most of their work. In fact, SCA operations are administered by NASA/Ellington, and one aircraft (911) is normally based at El Paso International Airport, Texas and the other (905) at Edwards. Modifications to permit the 747s to carry the 100-ton Orbiters include structural strengthening, the addition of large vertical fins at the tips of the tailplane for additional directional stability and support struts atop the upper fuselage

SCA No 1, tail number '905', was solely responsible for all Shuttle ferry flights until 1991. Formerly part of the American Airlines fleet, it also undertook a series of spectacular air-drops of the prototype Orbiter vehicle, *Enterprise*, during the Approach and Landing Test (ALT) programme of 1977. Here, '905' is pictured carrying *Discovery* back to Kennedy Space Center after the classified, Department of Defense-sponsored STS-33 mission in November 1989

Orbiters are lifted onto the SCAs using the Mate-DeMate Devices (MDDs) at Edwards and Kennedy Space Center. The 100-foot-high steel truss cantilevered structure is capable of precision positioning more than 220,000 pounds. It raises the Orbiter onto jacks for ferry flight servicing, then hoists it higher so that the SCA can be towed beneath the spacecraft in preparation for mating. An air-portable version of the MDD, known as the Stiff Legged Derrick, is available if the Orbiter lands at a site other than Edwards or Kennedy. This has only happened on one occasion – in 1982, when the STS-3/*Columbia* mission ended at the backup landing site at White Sands, New Mexico, after heavy rains had flooded Edwards

This is the No 2 SCA, tail number '911', seen undergoing post-flight checkouts after arriving at Edwards for the first time, in December 1990. Before being acquired by NASA and converted at Boeing's Wichita, Kansas plant, this aircraft served with Japan Airlines. It made its first Shuttle-carrying flight in May 1991, ferrying the latest (*Challenger*-replacement) Orbiter, *Endeavour*, from Rockwell International's Palmdale plant in California to Kennedy Space Center, Florida

(Dryden Flight Research Facility also operates one T-38A and two AT-38Bs for chase duties)

Splash!

Besides being accomplished test pilots, these men have one thing in common – they're all getting the traditional wetting that greets a homecomer after he's taken an aircraft out for the first time. Here, the aircraft is the Grumman X-29 forward-swept-wing demonstrator, and the flyers display a revealing range of reactions. Clockwise, from the top: repulsion from the USAF's Al Hoover, exultation from NASA's Ed Schneider; defiance from NASA's Bill Dana; and resignation from the US Navy's Ray Craig

NASA/Ellington

Ellington Field, formerly Ellington AFB, is located just south of the city of Houston, Texas and is operated as a 'reliever' airport for Houston Hobby. Nearby Johnson Space Center, home of NASA's manned space programme, is one of the flying tenent organsations at Ellington. Because it is situated close to where the astronauts live and work, most of Ellington's aircraft support their training and proficiency requirements

For many years, NASA employed this extraordinary aircraft — a much-modified Boeing YC-97J, the Aero Spacelines Super Guppy B-377 — to transport outsized loads. These have included satellites, the primary mirror for the Hubble Space Telescope, and the Shuttle Orbiter's streamlined tailcone, fitted to the spacecraft for ferry flights. Although operation of the Super Guppy was administered by personnel at Ellington Field, the aircraft itself was based at El Paso International Airport, Texas, where there was less airborne corrosive elements to tax its aging airframe. Alas, NASA has recently baulked at the $6–10 million cost of keeping the Super Guppy flying (it needs four new engines; T56s, as fitted to the French Guppy 201s). The whale-shaped aircraft was flown into long-term, storage at Davis-Monthan AFB, Arizona in July 1991, just as this book was being completed

The Super Guppy's flight crew pose with their pride and joy at Rockwell International's Palmdale, California plant. Super Guppy Project Pilot Frank Marlow is second from right. The outsized aircraft presented unique flying challenges to its crew – not least its tendency to mimic a wheelbarrow (the Super Guppy's nose-gear touches down first on landing, and leaves the runway last during takeoff – a characteristic inherited from its predecessors, the Boeing C-97 and B-377, which exhibited the same idiosyncrasy at light gross weights)

The Super Guppy's forward fuselage has been hinged across to accept an exotic outsized load — the Galileo planetary explorer and its Shuttle/Centaur upper-stage booster unit. Having been ferried with great care aboard the Super Guppy, Galileo is now safely *en route* to Jupiter!

Left Flying a succession of parabolic trajectories, NASA's Boeing KC-135 subjects trainee astronauts and scientific experiments (and experimenters) to short spells of near-weightlessness, or microgravity. About one-third of its passengers are physically sick on their first excursion, hence the aircraft's nickname, the *Vomit Comet*

Below A senior Kodak scientist in free float aboard the KC-135 whilst performing a fluid-physics experiment. Note the protective padding on the cabin interior, and the high-intensity lighting

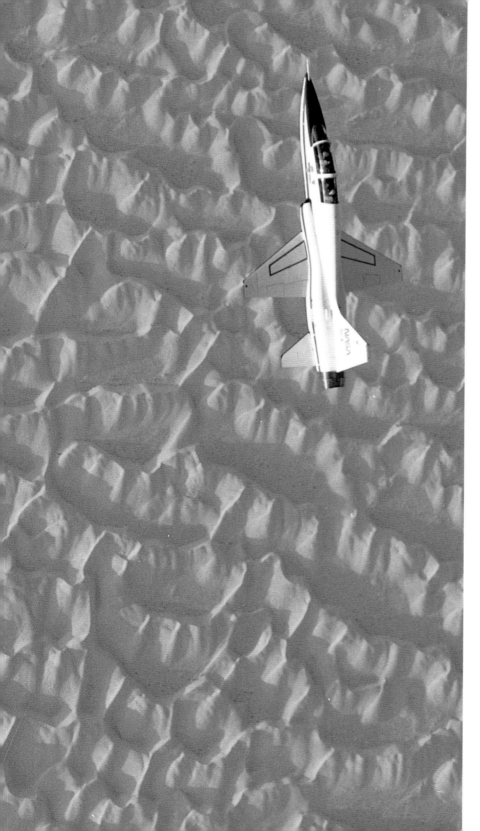

There are four Shuttle Training Aircraft, or STAs, on-strength at Ellington Field. STAs are heavily-modified Grumman Gulfstream 2 executive jets, equipped with highly-sophisticated computers and a flight control system capable of mimicking the Space Shuttle Orbiter in its very steep 'deadstick' (unpowered) approach and landing regime. In May 1991, a fifth Gulfstream 2 joined the fleet, but there were no immediate plans to convert it to STA standard. The STAs are often seen at the Shuttle landing sites at Kennedy Space Center, Florida, Edwards, California and White Sands, New Mexico – where one is pictured here accompanied by a T-38 chase-plane

Above The STA's cockpit has a split personality, being half-Gulfstream 2 bizjet (instructor's side) and half-Shuttle Orbiter (trainee's side). This enables astronauts to master the Shuttle's demanding flying characteristics, or simply maintain proficiency. Astronauts have been known to shoot as many as 900 practice landings in the STA prior to 'doing it for real' in the Shuttle. Here, veteran astronaut Brewster Shaw puts an STA through its paces

Left An STA basks in the sun at Ellington

NASA operates a sizable fleet of Northrop Talons, most of them out of Ellington Field. As recently as May/June 1991, the USAF presented the agency with a further six aircraft, all but one of them AT-38B models. The supersonic jets serve as spaceflight readiness trainers, and as chase-planes during Shuttle landings. In addition, the astronauts often use the T-38s to travel to distant engagements

Above Speed-brake extensions, sized as per the F-5, help the astronauts to perform simulated Shuttle landings

Right Posing power! This Shuttle crew (STS-41/*Discovery*) demonstrated their fondness for the T-38 by employing one as a backdrop to their official pre-flight portrait. A striking sunrise over Ellington Field completed the picture. Interestingly, this particular T-38 (NASA 915), was a ground instructional trainer at Holloman AFB, New Mexico for 12 years prior to NASA restoring it to flying status!

Two General Dynamics WB-57F aircraft facilitate high-altitude research, including the collection of tiny specks of meteorite material 'floating' in the upper atmosphere. Meteorological studies are also conducted. The WB-57F is an extended-wingspan derivative of Britain's classic English Electric Canberra jet bomber, built under licence in the United States. It totes a pair of auxiliary jet engines outboard of the main powerplants, and the NASA examples prickle with data-collection devices

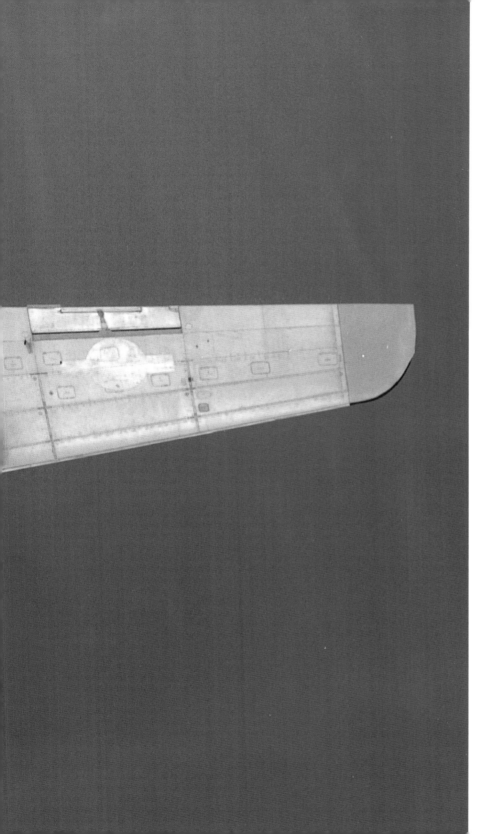

The WB-57F's huge wings glint in the sunlight. Note the evidence of previous USAF markings on both sides

(NASA/Ellington also operates a Grumman Gulfstream 1 for 'mission management' duties, such as taking Johnson Space Center senior management personnel to Florida for Orbiter Flight Readiness Reviews. In the immediate aftermath of the *Challenger* accident in January 1986, this aircraft was used to return the families of the deceased astronauts from Kennedy Space Center to Houston. In addition, NASA/Ellington administers operation of the two Boeing 747 Shuttle Carrier Aircraft, details of which appear in the 'Edwards' section)

NASA/Langley

Langley Research Center is located at Hampton, Virginia, and is primarily an R&D facility for advanced aerospace technology. One of its most impressive assets is the National Transonic Facility, a cryogenic windtunnel for high-Reynolds number research for subsonic and transonic flight regimes. In addition to its aeronautical activities, NASA/Langley manages several key space programmes, and was responsible for the Viking landers/orbiters which explored Mars in the late-1970s

Below NASA's sole Grumman Gulfstream 3 is, for reasons of protocol, the 'flagship of the fleet'. Flying under the callsign 'NASA 1', it transports the agency's top officials, including the NASA Administrator, to distant appointments. This aircraft is actually assigned to NASA Headquarters in Washington DC, but is operated out of Langley

Left NASA/Langley's aircraft fleet ranged across the ramp for inspection. Left to right: Bell 204B, Queen Air, F-106B Delta Dart, OH-58A Kiowa, F-5F Tiger II, Boeing 737 TSRV, Cessna 402B, T-38A Talon, PA-28 Arrow 4, Learjet Longhorn, Cessna U3A, and T-34C Turbo Mentor

NASA's Cessna 402B was built in 1972 and saw service as a 6/8-seat commuter/light-cargo aircraft before it came to Langley in May 1982. Since then, it has supported single-pilot IFR research, ride-quality studies and the Shuttle Exhaust Particle Experiment (SEPEX). The latter involved flying at 20,000 feet through the Shuttle's exhaust plume soon after launch, to collect particulate matter for laboratory analysis. NASA wished to discover if Shuttle launches influenced local climactic conditions. Particles were collected in silicon filters within nose-mounted probes, clearly visible here. Another interesting research programme involved flying through the smoke plumes from controlled forest-fires in Mexico's Yucatan peninsula, again to collect particulate samples for analysis

Beech T-34C Turbo Mentor '510' came to NASA/Langley direct from the factory in June 1978. In 1983 it facilitated laminar-flow glove experiments, and in 1985 and 1989 it joined the Cessna 402 in the SEPEX programme, flying at 4000 feet through the Shuttle exhaust plume soon after launch. Also in 1989, it served as a photo-platform in a vortex-detection experiment programme

Left Bell 204B '530' has served NASA at Langley since it came off the production line in 1964. It has been a smoke-sampling and aerial photography platform, and has supported tailboom strake aerodynamic research, but its most frequent duty is to air-drop free-flight scale models on remotely-piloted aerodynamic tests. Thus far, it has deployed scale models of the F-4, F-14, F-15, B-1, F-18, F-16XL, X-29 and X-31, the latter to evaluate high-alpha flight characteristics

Above right Langley's Learjet differs from the others in NASA's inventory by being a Longhorn derivative. Manufactured in 1972, it came to Langley in October 1984. It supported laminar-flow experiments from 1984–89 and is currently being used to measure electrical fields aloft and correlate these with meteorological conditions to determine conditions conducive to triggered lightning. This research might eventually enable rockets to be launched with confidence in conditions which are presently considered unsuitable

Right The exquisite lines of NASA's sole Northrop F-5F Tiger II are displayed to advantage high over the Virginia coastline. Manufactured in 1973 as an F-5E tactical fighter, it was converted to 'F' specification in 1974 and came to Langley in August 1989. It serves in a mission-support capacity

Above This prototype Piper PA-28RT-201 Arrow 4 was built in 1976 as a 200-series aircraft and was assigned to NASA two years later. Since that time, it has participated in stall-spin research, wingtip vortex turbine research, in-flight wake-vortex detection tests, and infrared 'off-surface' flow-visualisation studies

Left The largest aircraft at Langley is this Boeing 737, tail number '515', designated the Transport Systems Research Vehicle (TSRV). Built in 1968, it has the distinction of being the first production Boeing 737. It has supported efforts to pioneer a vast range of civil aviation technological innovations during its 17-year period with NASA, evaluating and demonstrating an experimental Microwave Landing System (MLS), 'glass' (CRT) cockpit displays, helmet-mounted displays for precision manual landings, and subsequently, a ground-air data link in lieu of voice communication, a wind-shear detection system and satellite-based (GPS) precision navigation systems for automated landings

Left Manufactured in 1976, NASA's sole Beech 65-B80 Queen Air came to Langley the following year. The ten-seat utility aircraft fulfils mission-support duties

Right An old-timer, still going strong. NASA/Langley's Cessna U3A, tail number '505', came off the production line in 1957. The five-seat utility aircraft was assigned to an engine slipstream study programme in 1984, but today undertakes mission-support activities. The U3A is the military version of the Cessna 310

Below NASA/Langley's Bell OH-58A Kiowa serves in a mission-support capacity

Although not strictly eligible for inclusion, as it was withdrawn from service in 1990, Langley's trusty Convair F-106B Delta Dart really does deserve a place. Built in 1957, it entered the NASA inventory in January 1979. From then until 1986 it undertook vortex and storm-hazard research, being flown into tempestuous weather systems and sustaining over 700 lightning strikes in the cause of air safety. The F-106 was subsequently fitted with leading-edge vortex flaps (pictured) and put through an extensive flight test programme to validate their effect on performance. It was the world's last airworthy F-106

(NASA/Langley also operates one Northrop T-38A and one AT-38B on chase and other mission support duties, plus an F-16A for similar tasks, and pilot proficiency training for possible future flight experiments)

NASA/Lewis

Lewis Research Center is situated in Cleveland, Ohio. It is NASA's lead centre for R&D on aircraft propulsion, space propulsion, space power-generation and satellite communications

Looking pristine in its red, white and blue livery is the Lewis-based Grumman Gulfstream 1, callsign 'NASA 5'. NASA operates a total of four Gulfstream 1s from four separate centres

Right Researchers at NASA/Lewis use this de Havilland DHC-6 Twin Otter to conduct in-flight ice-accretion studies, cloud-physics research, ice protection system development, and aerodynamic performance and aircraft stability/control studies. The Twin Otter came to Lewis in 1981 and was modified at a cost of about half a million dollars

One of the most distinctive-looking aircraft in NASA service is this heavily-instrumented Rockwell OV-10A Bronco, which has retained its military markings. It was acquired in 1984 to engage in fundamental aero-acoustic research for the ATP programme, but now provides flight test support for a multitude of research programmes. One of the most attractive features of the Bronco from the research standpoint is its wide flight envelope. Its airspeed range, for example, is 55 to 350 knots in addition, it is relatively inexpensive to operate. The aircraft has an exceptional capability for carrying pilotised flight instrumentation and is currently assigned to a joint NASA/General Electric programme investigating active propeller noise-cancellation technology

(NASA/Lewis also operates a Beech T-34B in support of environmental, microgravity and physiological educational programmes for secondary school students, and a Learjet 25 for aeronautical, microgravity and other research activities)

NASA/ Wallops

Wallops Flight Facility is located at Wallops Island on Virginia's east coast, and is administered by NASA's Goddard Space Flight Center in Greenbelt, Maryland. It became America's third orbital launch site in February 1961, when the Explorer 9 balloon-satellite was lofted atop a Vought Scout. Since 1945, over 13,000 suborbital sounding-rockets have been launched from the pads on Wallops Island. Wallops has three runways, ranging in length from 4000 to nearly 9000 feet. More than 300 aeronautical research missions are logged annually, together with dozens of range-support flights. The Microwave Landing System (MLS), which has become the operational standard at airports around the world, was developed and tested at Wallops. Another interesting venture has been the development of an experimental high-speed turnoff ramp, which is expected to enable aircraft at busy airports to exit the runway immediately upon landing. This will allow closer spacing of incoming aircraft. The high-speed turnoff includes a buried magnetic cable which will automatically signal the steering mechanisms in the aircraft to make the turn without pilot input

Wallops Flight Facility's UH-1, Skyvan, Electra and Orion receiving attention in the NASA hangar

Left NASA's T-39 Sabreliner was built in-house from the pieces of two T-39s that had been retired from the USAF and US Navy. The sleek twinjet provides scientists with a fast (670 mph), high-altitude (40,000 feet) instrument platform

Below left NASA operates two Bell UH-1s out of Wallops Island. They used to perform water recoveries of rocket-launched payloads, but are now employed as remote-sensing platforms for vegetation and forestry studies, and undertake air-sea rescue operations during aeronautical research flights

Right This Shorts SC-7 Skyvan has several primary tasks, including deft mid-air retrievals of sounding-rocket payloads as they descend by parachute from the upper reaches of the atmosphere or from space itself. That role, coupled with the aircraft's unfortunate physical attributes, have earned it an interesting nickname – the *Ugly Hooker*. For the scientific community, sounding-rockets commonly 'fill the gap' between the maximum altitude for balloons (about 30 miles) and the minimum altitude for orbiting satellites (about 100 miles). Since 1979, the Skyvan has captured 99 per cent of the payloads it has been despatched to recover

Right The acquisition, in 1977 and '78, of two four-engined Lockheed propjets — the NP-3A Orion and the L-188 Electra — greatly expanded Wallops Flight Facility's capability, with speed increases up to 350 mph and altitude up to 25,000 feet, as well as better dependability and safety of operations. The Orion is a highly-efficient instrument platform for scientific research projects. It incorporates a 'walk-on' approach for various user-owned instrument systems, which easily integrate with the airplane's electrical power system, timing apparatus, voice circuits and video-recording facilities and navigation systems. When not performing range-surveillance duties during sounding-rocket flight operations, the Orion serves that sector of the scientific community concerned with the earth and oceans, measuring everything from tree heights to soil moisture-content and oceanic wave patterns. It has contributed to ice-cap and sea-ice studies in Greenland, ocean wave studies during Icelandic storms, and tree-mapping over the rain-forests of Costa Rica. NASA/Wallops now operates a second Orion, a P-3B variant which has retained its US Navy livery

(NASA/Wallops also operates a Beech King Air 200, mainly on administrative duties)

Inset NASA's Lockheed Electra also incorporates a 'walk-on' approach to experiments. It is a key atmospheric research platform designed to incorporate state-of-the-art instruments, such as lidar systems for measuring aerosols in the stratosphere. It has supported such diverse activities as studies of the El Chichon volcanic eruption in Mexico in 1982 and its global effects, stratospheric aerosol surveys in support of the SAM-2 satellite programme from the North Pole to the southern tip of Chile, and the Global Tropospheric Experiment (GTE), a programme to study the worldwide dynamics of the troposphere. In the summer of 1991, the Electra again studied volcanic effects. The massive outpouring of volcanic materials from Mount Pinatubo in the Philippines – believed to be double that from any eruption in the last century – prompted NASA to assemble a quick-response research team to help evaluate the global atmospheric effects of this catastrophic event. Scientists collected vast quantities of data as the Electra, temporarily based at Grantley Adams Airport on Barbados, took samples from the giant pillars of volcanic smoke and gas being blown towards the Caribbean

NASA/Elsewhere

There are several aircraft at the other NASA centres: Kennedy Space Center, near Titusville, Florida operates three Bell UH-1 helicopters and a Grumman Gulfstream 1; Stennis Space Center at Bay St Louis, Mississippi, where the Shuttle liquid-propellant main engines are tested, has a lone Learjet 23; NASA's Marshall Space Flight Center at Huntsville, Alabama – home of many of NASA's most important programmes – operates a Grumman Gulfstream 1 and a Beech King Air 200; and the Jet Propulsion Laboratory (JPL), at Pasadena near Los Angeles – from which most of NASA's spectacular planetary missions are managed – also has a King Air 200. JPL is operated for NASA by the California Institute of Technology, or Caltech

One of the three Bell UH-1s which currently operate out of NASA's Kennedy Space Center in Florida, where Shuttle missions not only launch, but often land – on the Shuttle Landing Facility (SLF), located about four miles from the two Shuttle launch pads. Here, a UH-1 prepares to airlift an 'injured' member of the launch pad close-out crew to a local hospital during a simulated emergency evacuation exercise involving the Orbiter *Atlantis*

Inventory (at summer 1991)

RD – Research and Development
PS – Programme Support
    SFT – Astronaut Space Flight Training
    STA – Shuttle Training Aircraft
    SCA – Shuttle Carrier Aircraft
MM – Mission Management

NASA/Ames Research Center, Moffett Field, California

Aircraft	Serial	Type	Owner
Bell XV-15	703	RD	NASA
BAe YAV-8B Harrier	704	RD	US Navy
Boeing/DHC QSRA	715	RD	NASA
Sikorsky UH-60 Blackhawk	748	RD	US Army
Sikorsky UH-60 Blackhawk	750	RD	US Army
Bell NAH-1S	736	RD	US Army
BAe AV-8C Harrier	719	PS	US Navy
Lockheed YO-3A	718	RD	NASA
Bell UH-1H Huey	734	RD	US Army
Beech King Air 200	701	PS	NASA
Sikorsky SH-3G	735	RD	US Navy
Northrop T-38A Talon	722	PS	NASA
Bell XV-15	702	RD	NASA*
McD DC-8-72	717	RD	NASA
Lockheed C-141 'Kuiper'	714	RD	NASA
Lockheed C-130	707	RD	NASA
Gates Learjet 24	705	PS	NASA
Lockheed ER-2	706	RD	NASA
Lockheed ER-2	709	RD	NASA
Lockheed TR-1	708	RD	USAF
Northrop T-38A Talon	61-870	PS	NASA

* On loan to Bell, Forth Worth, Texas

NASA/Dryden Flight Research Facility, Edwards, California

Aircraft	Serial	Type	Owner
PIK 20E sailplane	803	RD	NASA
General Dynamics F-16XL	849	RD	USAF
General Dynamics F-16XL	846	RD	USAF
General Dynamics F-16 AFTI	750	RD	USAF
Lockheed F-104 Starfighter	826	RD	NASA
Lockheed TF-104 Starfighter	825	RD	NASA
Grumman X-29	049	RD	DARPA
McD F-15 Eagle HiDEC	835	RD	USAF
Boeing NB-52	008	RD	USAF
McD F-18 HARV	840	RD	US Navy
Convair 990	810	RD	NASA
Piper PA-30 Twin Comanche	808	PS	NASA
Northrop T-38A Talon	821	PS	NASA
McD F/A-18 Hornet	841	PS	US Navy
McD F/A-18 Hornet	842	PS	US Navy
McD F/A-18 Hornet	843	PS	US Navy
McD TF-18 Hornet	845	PS	US Navy
McD F/A-18 Hornet	847	PS	US Navy
McD TF-18 Hornet	848	PS	US Navy
Lockheed SR-71A	832	RD	USAF
Lockheed SR-71A	844	RD	USAF
Lockheed SR-71B	831	RD	USAF
Northrop AT-38B Talon	62-3715	PS	NASA
Northrop AT-38B Talon	62-3746	PS	NASA

Continued overleaf

NASA/Ellington Field (JSC), Houston, Texas

Aircraft	Serial	Type	Owner
Northrop T-38A Talon	901	PS/SFT	NASA
Northrop T-38A Talon	902	PS/SFT	NASA
Northrop T-38A Talon	903	PS/SFT	NASA
Northrop T-38A Talon	904	PS/SFT	NASA
Northrop T-38A Talon	906	PS/SFT	NASA
Northrop T-38A Talon	907	PS/SFT	NASA
Northrop T-38A Talon	908	PS/SFT	NASA
Northrop T-38A Talon	909	PS/SFT	NASA
Northrop T-38A Talon	910	PS/SFT	NASA
Northrop T-38A Talon	912	PS/SFT	NASA
Northrop T-38A Talon	913	PS/SFT	NASA
Northrop T-38A Talon	914	PS/SFT	NASA
Northrop T-38A Talon	915	PS/SFT	NASA
Northrop T-38A Talon	916	PS/SFT	NASA
Northrop T-38A Talon	917	PS/SFT	NASA
Northrop T-38A Talon	918	PS/SFT	NASA
Northrop T-38A Talon	919	PS/SFT	NASA
Northrop T-38A Talon	920	PS/SFT	NASA
Northrop T-38A Talon	921	PS/SFT	NASA
Northrop T-38A Talon	923	PS/SFT	NASA
Northrop T-38A Talon	924	PS/SFT	NASA
Northrop T-38A Talon	955	PS/SFT	NASA
Northrop T-38A Talon	956	PS/SFT	NASA
Northrop T-38A Talon	960	PS/SFT	NASA
Northrop T-38A Talon	961	PS/SFT	NASA
Northrop T-38A Talon	962	PS/SFT	NASA
Northrop T-38A Talon	963	PS/SFT	NASA
Northrop AT-38B Talon	964	PS/SFT	NASA
Northrop AT-38B Talon	68-8133	PS/SFT	NASA
Northrop AT-38B Talon	65-10450	PS/SFT	NASA
Northrop AT-38B Talon	68-8116	PS/SFT	NASA
Northrop AT-38B Talon	61-0907	PS/SFT	NASA
Northrop AT-38B Talon	64-13267	PS/SFT	NASA
General Dynamics WB-57F	926	PS	USAF
General Dynamics WB-57F	928	PS	USAF
Boeing KC-135	930	PS	NASA
Grumman G-1159 Gulfstream 2	944	PS/STA	NASA
Grumman G-1159 Gulfstream 2	945	PS/STA	NASA
Grumman G-1159 Gulfstream 2	946	PS/STA	NASA
Grumman G-1159 Gulfstream 2	947	PS/STA	NASA
Grumman G-1159 Gulfstream 2	948	PS	NASA
Boeing 747-123	905	PS/SCA	NASA
Boeing 747-146	911	PS/SCA	NASA
AeS Super Guppy B-377	940	PS	NASA**
Grumman Gulfstream 1 G-159	2NA	MM	NASA

** Withdrawn from service in July 1991

Stennis Space Center at Bay St Louis, Mississippi operates just one aircraft, this Learjet 23

NASA/Langley Research Center, Hampton, Virginia

Piper PA-28 Arrow	519	RD	NASA
Boeing 737-100	515	RD	NASA
Gates Learjet 28/29	566	RD	NASA
Cessna 402B	503	RD	NASA
Cessna U3A	505	PS	NASA
Beech Queen Air B80	506	PS	NASA
Northrop T-38A Talon	511	PS	NASA
Northrop F-5F Tiger II	550	PS	USAF
Beech T-34C Mentor	510	PS	NASA
Bell 204B	530	PS	NASA
Bell OH-58A Kiowa	540	PS	US Army
Grumman Gulfstream 3	1NA	MM	NASA
Northrop AT-38B Talon	63-8117	PS	NASA
General Dynamics F-16A	516	RD	NASA

NASA/Lewis Research Center, Cleveland, Ohio

Gates Learjet 25	616	PS	NASA
DHC-6 Twin Otter	607	RD	NASA
Rockwell OV-10A Bronco	636	RD	US Navy
Beech T-34B	614	PS	NASA
Grumman Gulfstream 1 G-159	5NA	MM	NASA

NASA/Kennedy Space Center, near Titusville, Florida

Bell UH-1M	417	PS	NASA
Bell UH-1M	418	PS	NASA
Bell UH-1H	419	PS	NASA
Grumman Gulfstream 1 G-159	4NA	MM	NASA

NASA/Marshall Space Flight Center, Huntsville, Alabama

Grumman Gulfstream 1 G-159	3NA	MM	NASA
Beech King Air 200	9NA	MM	NASA

NASA/Stennis Space Center, Bay St. Louis, Mississippi

Gates Learjet 23	933	PS	NASA

NASA/Wallops Flight Facility, Wallops Island, Virginia

Bell UH-1B	424	PS	NASA
Bell UH-1H	415	PS	NASA
Shorts SC-7 Skyvan	430	PS	NASA
NA T-39A Sabreliner	431	PS	NASA
Lockheed NP-3A Orion	428	PS	NASA
Lockheed P-3B Orion	152735	PS	NASA
Lockheed L-188 Electra	429	PS	NASA
Beech King Air 200	8NA	MM	NASA

NASA/Jet Propulsion Laboratory, Pasadena, California

Beech King Air 200	7NA	MM	NASA